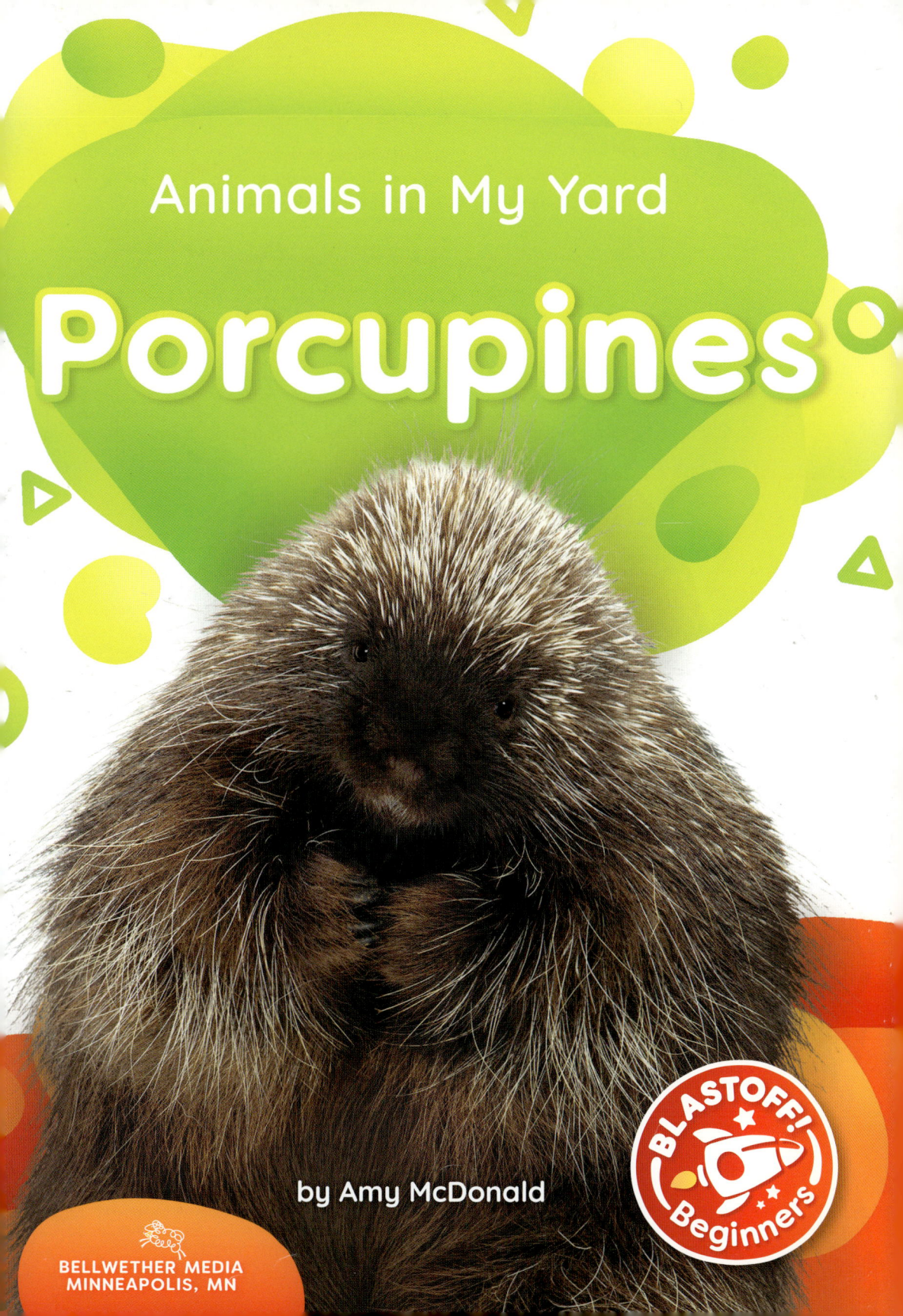

Blastoff! Beginners are developed by literacy experts and educators to meet the needs of early readers. These engaging informational texts support young children as they begin reading about their world. Through simple language and high frequency words paired with crisp, colorful photos, Blastoff! Beginners launch young readers into the universe of independent reading.

Sight Words in This Book

a	eat	like	their
and	find	long	them
are	funny	out	they
at	have	sit	this
can	in	that	to
come	is	the	what

This edition first published in 2021 by Bellwether Media, Inc.

No part of this publication may be reproduced in whole or in part without written permission of the publisher. For information regarding permission, write to Bellwether Media, Inc., Attention: Permissions Department, 6012 Blue Circle Drive, Minnetonka, MN 55343.

Library of Congress Cataloging-in-Publication Data

Names: McDonald, Amy, author.
Title: Porcupines / by Amy McDonald.
Description: Minneapolis, MN : Bellwether Media, 2021. | Series: Blastoff! beginners : Animals in my yard | Includes bibliographical references and index. | Audience: Ages PreK-2 | Audience: Grades K-1 |
Identifiers: LCCN 2020029478 (print) | LCCN 2020029479 (ebook) | ISBN 9781644873618 (library binding) | ISBN 9781648340628 (ebook)
Subjects: LCSH: Porcupines--Juvenile literature.
Classification: LCC QL737.R652 M395 2021 (print) | LCC QL737.R652 (ebook) | DDC 599.35/97--dc23
LC record available at https://lccn.loc.gov/2020029478
LC ebook record available at https://lccn.loc.gov/2020029479

Text copyright © 2021 by Bellwether Media, Inc. BLASTOFF! BEGINNERS and associated logos are trademarks and/or registered trademarks of Bellwether Media, Inc.

Editor: Christina Leaf   Designer: Jeffrey Kollock

Printed in the United States of America, North Mankato, MN.

Table of Contents

Porcupines!	4
Body Parts	6
The Lives of Porcupines	12
Porcupine Facts	22
Glossary	23
To Learn More	24
Index	24

Porcupines!

What is that funny animal?
A porcupine!

Body Parts

Porcupines have **quills**. They are sharp.

quills

They have long teeth. They like to chew.

teeth

They have claws.
They can climb.

claws

The Lives of Porcupines

Porcupines live in forests. They sit in trees.

They sleep in **dens**.
They come out at night.

den

They find food after dark. They eat bark, leaves, and seeds.

bark

leaves

seeds

Porcupines stay safe. Quills **protect** them.

The animals **rattle** their quills. This is a warning. Stay back!

Porcupine Facts

Porcupine Body Parts

quills

teeth

claws

Porcupine Food

bark

leaves

seeds

Glossary

dens

homes for porcupines

protect

to keep safe from harm

quills

long, sharp hairs on a porcupine's body

rattle

to shake to make noise

To Learn More

ON THE WEB

FACTSURFER

Factsurfer.com gives you a safe, fun way to find more information.

1. Go to www.factsurfer.com.

2. Enter "porcupines" into the search box and click 🔍.

3. Select your book cover to see a list of related content.

Index

bark, 16	protect, 18	warning, 20
chew, 8	quills, 6, 18, 20	
claws, 10, 11	rattle, 20	
climb, 10	seeds, 16, 17	
dens, 14, 15	sharp, 6	
food, 16	sit, 12	
forests, 12	sleep, 14	
leaves, 16, 17	teeth, 8	
night, 14	trees, 12	

The images in this book are reproduced through the courtesy of: Eric Isselee, front cover, pp. 3, 4, 5, 10, 22; Sasha Samardzija, p. 6; Coulanges, pp. 6-7; Jim Cumming/ Alamy, p. 8; Warren Metcalf, pp. 8-9; Jeff Caverly, pp. 10-11; Japan's Fireworks, pp. 12-13; MBoe, pp. 14-15; Tony Rix, pp. 16-17; Gregory Johnston, p. 16; schankz, p. 17 (leaves); Auhustsinovich, p. 17 (seeds); Debbie Steinhausser/ Alamy, pp. 18-19; Geoffrey Kuchera, pp. 20-21, 22 (seeds); Nancy Bauer, p. 22 (bark); Creatista, p. 22 (leaves); Tawin Mukdharakosa, p. 23 (dens); Scenic Shutterbug, p. 23 (protects); Michal Ninger, p. 23 (quills); Jukka Jantunen, p. 23 (rattle).